LADYBIRDS & LOBSTERS SCORPIONS & CENTIPEDES...

Natural History Museum Publications
London

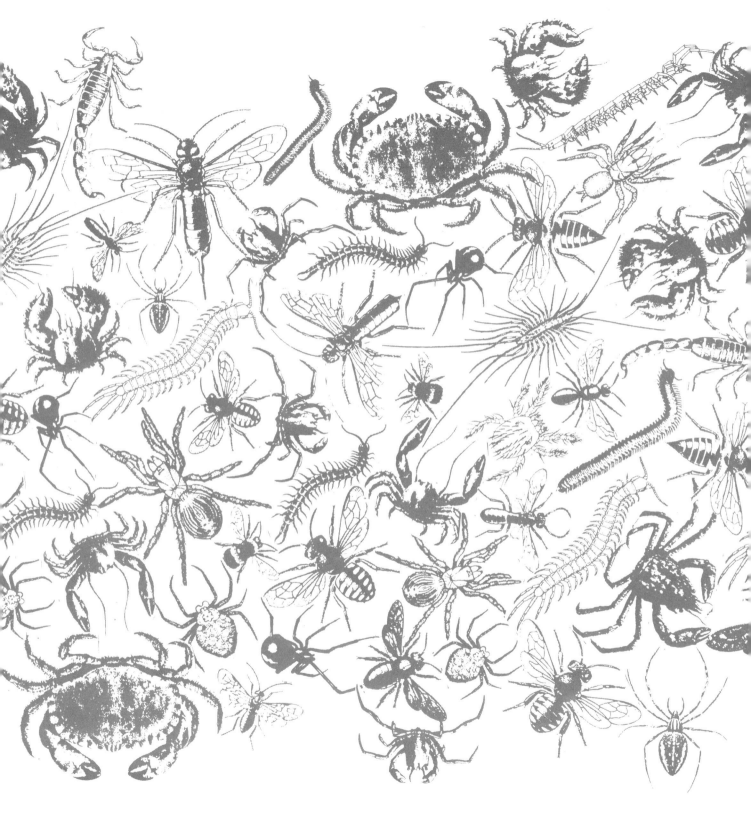

Contents

Preface — 5

Chapter 1
Introducing arthropods — 6

Chapter 2
How arthropods grow — 16

Chapter 3
All the colours of the rainbow — 28

Chapter 4
Battling for survival — 42

Chapter 5
Food and feeding — 50

Chapter 6
Masterbuilders — 58

Chapter 7
Taking up residence — 68

Chapter 8
Arthropods and humans — 75

Chapter 9
Ancient arthropods — 84

Further reading — 103

Acknowledgements and index — 104

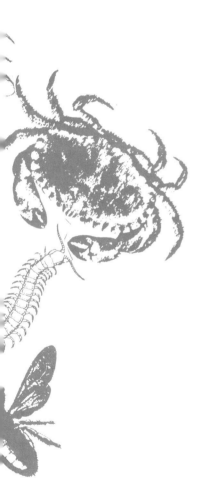

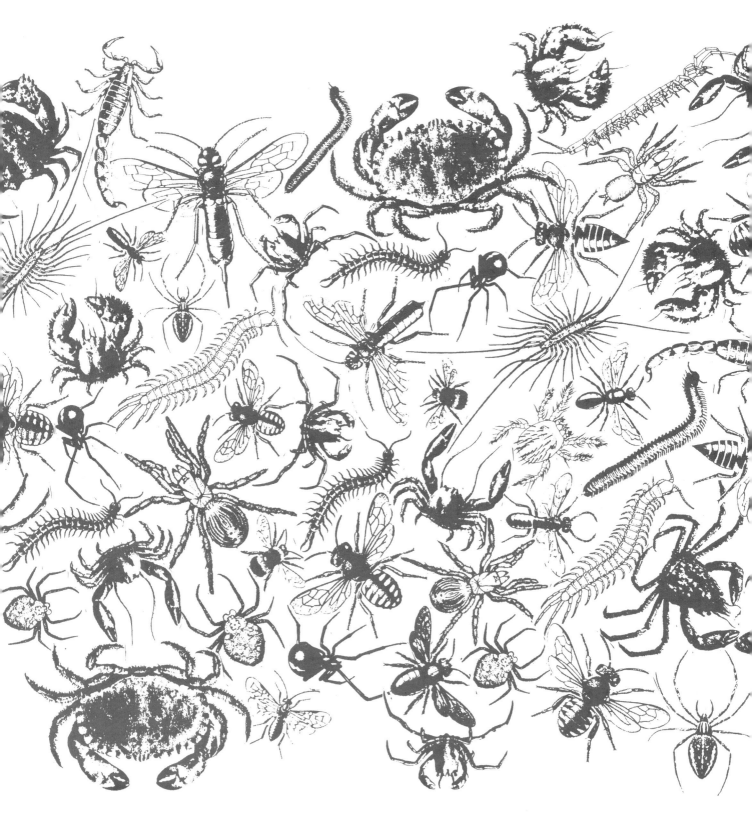

Preface

The arthropod group contains animals as diverse as ladybirds, lobsters, scorpions and centipedes. Although these animals may seem to be very different from each other, they all have two striking things in common: their limbs have joints and their skeletons are on the outside of the body.

This book looks at a vast range of arthropods and highlights their versatility. By examining the ways in which they are adapted to survive in a wide range of frequently hostile environments, the book goes some way towards explaining the enormous success of arthropods.

It describes, among other things, their colourful bodies, which help to fool prey and predators, and the complex structures that some build as communal homes.

Inevitably the ubiquitous arthropods come into contact with humans. The book examines our relationships with arthropods, from harmless house spiders to the mosquitoes that transmit malaria and the locusts that destroy our crops.

The origins of the group are also considered. The fossil evidence has much to reveal about the ways in which arthropods have evolved and adapted to new environments over millions of years.

The book is a companion to '*Creepy Crawlies*' –*an exhibition about arthropods,* which opened at The Natural History Museum in 1989. Both the book and the exhibition were produced with the help of a large number of people. I should like to thank the writers, scientists and designers who have contributed to these projects.

NEIL CHALMERS Director
The Natural History Museum
1991

Chapter 1
Introducing arthropods

What is an arthropod?

The term 'arthropod', meaning 'jointed limb', was first coined over 140 years ago. Today it is used to describe a very large group of animals found all over the world.

How many kinds are there?

So far around 1.3 million different kinds of arthropods have been found but there are certainly a lot more species to be discovered. Some scientists believe there may be as many as 30 million altogether.

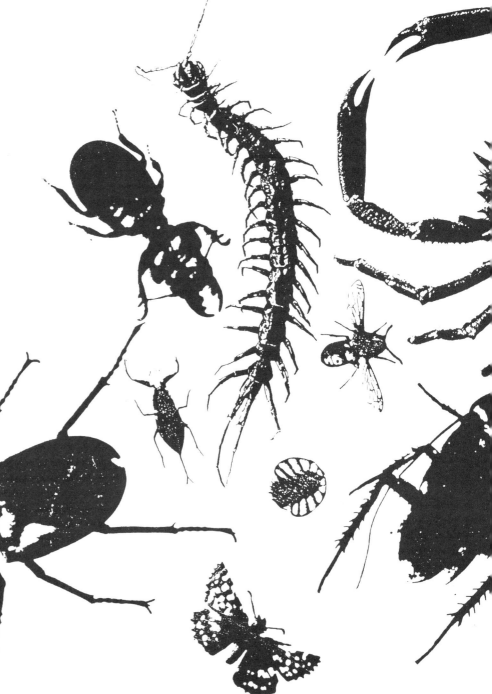

How long have arthropods existed?

Arthropods have a spectacularly long history and were around long before most of the better known animal groups. The earliest known arthropods have been found in rocks over 600 million years old. At that time all arthropods were marine, but gradually land-living forms evolved. By 300 million years ago one group, the insects, had taken to the air and enormous dragonflies darted around the primeval forests. Today, insects are the largest and most diverse arthropod group and still the only one able to fly.

The giant dragonfly had a wingspan of up to 80 centimetres

What makes an animal an arthropod?

All arthropods have certain characteristics in common. They have jointed limbs which nearly always occur in pairs. These can vary in number from three pairs in most insects to over 350 pairs in some millipedes.

Probably the most important single characteristic common to virtually all arthropods is their hard outer skeleton, called an exoskeleton. In most arthropods the exoskeleton incorporates material called chitin. Depending on the balance of its constituents, the exoskeleton may be rigid or flexible, waterproof or permeable.

One disadvantage of an exoskeleton is that it will not expand. To grow, arthropods must shed their old skeleton for a new one which, before it hardens, is sufficiently flexible to allow some expansion.

Having a tough skeleton outside the body does, however, have many advantages. In land-living arthropods the waterproof exoskeleton protects them from the moisture-sapping rays of the sun. This factor, more than any other, has enabled arthropods to colonize every terrestrial habitat, including some of the world's hottest and driest places.

But their success has not been limited to the land. In the sea, where being watertight is important for maintaining the correct concentration of body fluids, crustaceans are the dominant group. Arthropods, such as krill and microscopic copepods, occur in enormous numbers and form the basis of many important marine food chains.

Arthropods are important for our survival. On land, their activities as pollinators provide us with many plant products, including fruits, and some marine arthropods are a direct food source. But they are also our competitors for the Earth's resources – undoubtedly we have a constant battle trying to prevent them from consuming our food crops. Their ability to colonize so many different environments, to adapt rapidly to changes in their habitats and to utilize a wide range of substances for food has made arthropods the most successful and diverse animals on Earth.

Arthropods are usually divided into four main groups:

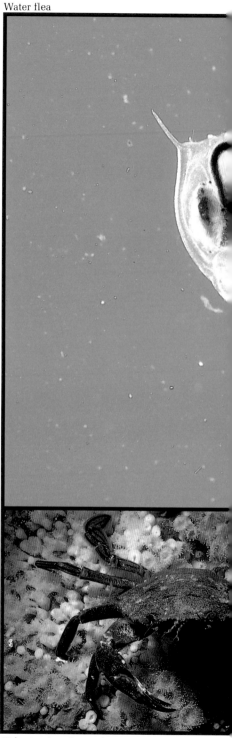

Water flea

Velvet swimming crab

Common lobster

Crustaceans – crabs and their relatives

Crustaceans, which include some of the largest arthropods known, are predominantly aquatic and occur throughout the world in both marine and freshwater environments. They vary greatly in size – some water fleas are only 0.25 millimetres long while the largest crabs have a body width of 30 centimetres and a leg span of some 3 metres.

Red coral shrimp

Woodlouse

Edible crab

Giant centipede   Stone centipede

Myriapods – centipedes and millipedes

Myriapods fall into two distinct groups, the centipedes (Chilopoda) and millipedes (Diplopoda).

There are about 1800 species of centipedes. Most of them are active predators with poisonous fangs. They vary in length from a few millimetres to more than 20 centimetres.

Millipedes species number about 7000, all of which are vegetarian, feeding mainly on decaying plant material. The smallest species are only a couple of millimetres in length; the longest up to 30 centimetres.

Stone centipede

Soil mites

Harvestman spider

Woodlouse spider

Arachnids – spiders and their relatives

Spiders, scorpions, mites, ticks and eight other smaller groups make up the arachnids. All members of the group, with the exception of some mites, have four pairs of legs. Both spiders and scorpions are predatory and often poisonous. Altogether there are some 75 000 species of arachnids of which about half are spiders.

Desert scorpion

Imperial scorpion

Lace-web spider

Desert locust

Longhorn beetle larva

Insects

For sheer numbers and diversity insects are unsurpassed by any other animal group. Around 1.1 million species are known and there are probably millions more awaiting discovery. Insects are characterized by having three distinct body parts (head, thorax and abdomen), three pairs of legs and usually two pairs of wings.

With the exception of the sea, practically every type of environment can support insect life.

Lesser stag beetle

Peacock caterpillars

Common earwig

Lace wing

Holly Blue butterfly

Burying beetle

ng mantis

Cockchafer

Common wasp

Blue Mormon butterfly

15

Chapter 2
How arthropods grow

All arthropods live within the confines of a rigid exoskeleton. Although the inner organs expand as arthropods mature, the outer skeleton remains fixed in size and shape.

How then do arthropods manage to grow?

If you lived inside a suit of armour, you too would have to shed it too grow!

To increase in size, an arthropod has to discard its tight-fitting cuticle, and replace it with a more spacious new one. This process is called moulting.

Moulting

At a certain time hormones in an arthropod's body begin to act upon its outer covering, which soon becomes thin and weak. Often after a period of vigorous activity the brittle cuticle becomes loosened and finally splits. Underneath, a new skeleton has been forming. Now the arthropod pulls itself free of the old dried-out skeleton to reveal the fresh one below. At first the new cuticle is soft, pale and limp, but soon it expands, hardens and becomes coloured, giving the arthropod a new protective coating.

Centipedes, millipedes, crustaceans and some arachnids moult at certain intervals throughout their lives. In contrast, most insects moult only during the early part of their lives, reaching sexual maturity with their last moult.

Some arthropod larvae look like perfect miniatures of their parents – they simply increase in size as they moult and grow. Others differ from the adults in almost every respect – they change dramatically as they grow.

The development of insects

As insects grow, most undergo an incredible transformation – there is often no resemblance at all between the larvae and the adult insects.

This transformation is illustrated by the mayfly and its larva.

Would you have guessed that these two creatures were both phases in the life of a mayfly?

Mayfly larvae (or nymphs) resemble shrimps. They live underwater in ponds or rivers and eat algae and decaying plant material. Some kinds moult more than 20 times while in their aquatic environment. At a particular time of year when they have reached a certain stage of growth, they rise to the surface and moult to reveal the penultimate stage in their lives. Although these winged creatures resemble adults, their legs are not fully developed. They fly off to find a sheltered place, and moult for the last time, revealing the mature adults.

Another example of diversity in a life cycle is the nymphalid butterfly and its larva.

The larvae of these butterflies (caterpillars) finally reach their full size after months or even years of almost non-stop eating. Then they begin to prepare themselves for the final stage in their development. They spin silk

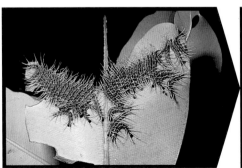

pads to attach themselves securely to twigs or plant stems. In this suspended state their bodies begin to change. (The transitional form is known as the pupa.) When the transformation is complete the winged adult emerges.

Why do many insects go through these amazingly different forms as they grow? Why don't they emerge from the egg as tiny winged insects?

One reason is that in insects, larval and adults stages fulfil very specific functions. The prime function of larvae is to eat to provide the energy for growth. The main function of the adult is to reproduce. An extreme example of this is the adult mayfly which may live only for a day – just long enough to find a mate and for the female to lay her eggs!

Another reason is that by having different habitats and food sources, insect larvae and adults do not compete with each other.

A Coxcomb Prominent caterpillar eats its recently shed skin

Although they are superb eating machines with large efficient jaws, many insect larvae are poorly adapted for other tasks:

- they cannot mate since their sexual organs are undeveloped;

- they are slow, clumsy movers since their legs are rudimentary and they have no wings;

- they cannot withstand attacks since they are covered only by a fairly thin cuticle. (The really hard exoskeleton develops only in the adult after the last moult.)

Because they are physically vulnerable, larvae are often camouflaged to evade predators; or they may have bright warning colours to show that they are poisonous to eat. (You can find out more about arthropod colours in chapter 5.)

The development of crustaceans

Freshwater and marine crustaceans develop differently. Freshwater crustaceans tend to follow a direct line of development – the young being small replicas of the adults. Marine crustaceans, on the other hand, go through dramatic changes as they grow.

It is likely that, as for insects, the development paths were originally determined partly by the availability of food. The sea has a variety of food sources, and marine crustaceans may have evolved so that the larvae fed on a different food source from the adults.

A good example of this among marine crustaceans is the edible crab.

The adult edible crab and its larval form are about as different as you could imagine.

The adult lives among rocks and crevices in coastal waters. It is carnivorous and feeds on mussels, breaking open their shells with its powerful pincers. In contrast, the comma-shaped early larval stage floats freely in the water. It catches prey by lashing its forked tail through swarms of plankton, some of which becomes trapped between the tail forks and the larva's shell. The food is then manoeuvred towards the larva's mouth.

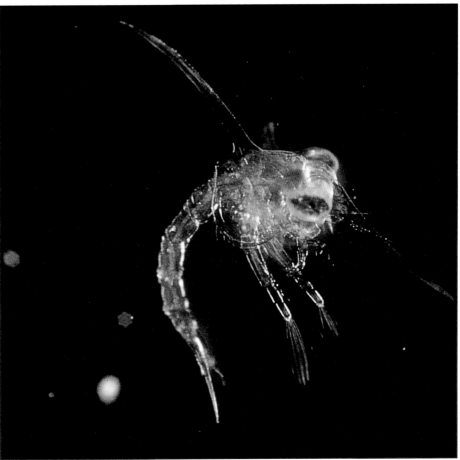

Adult barnacles bear little resemblance to their larvae. Adults spend their lives glued to rocks or the bottoms of boats. Early planktonic larval stages float freely while more developed ones crawl about on the seabed. Using their feathery limbs, the adults filter the water for plankton of almost any shape or size, while the larvae use their hairy antennae to sweep the water for smaller planktonic plants. Clumps of these plants caught on the larva's sharp hairs are fed back through the mouthparts.

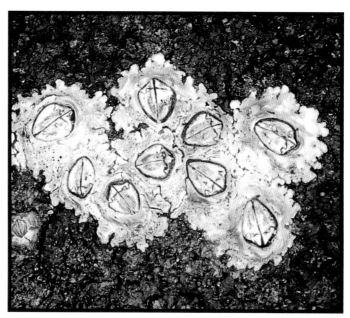

Freshwater crustaceans have adapted to an environment which has a limited range of available food. Both adults and young exploit the same food source.

The adult and larval forms of the freshwater crayfish resemble each other quite closely. Both are found in clean lowland rivers and feed by scavenging.

The development of myriapods

Millipedes

Rather than simply expanding in size, millipedes grow by adding groups of three or more segments after each moult.

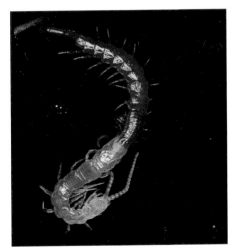

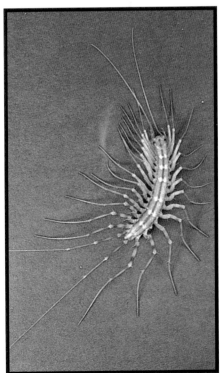

Centipedes

Centipedes can be divided into two groups according to the way they develop - one group hatches from the egg with a full complement of legs, (left) while the other hatches with seven pairs of legs or fewer and adds segments and legs in the early stages of development (right).

The development of arachnids

Young spiders (spiderlings) closely resemble their parents. Each time they moult they grow larger and become more highly developed.

An interesting example of moulting is found among tarantulas. They stop feeding and withdraw into retreats to moult. There they lie on their backs and the lengthy process begins.

The young tarantula shown below lack the density of colour and hair of the adult (inset). Since they are likely to fall prey to other spiders – even their own parents – spiderlings have to become independent as soon as possibl

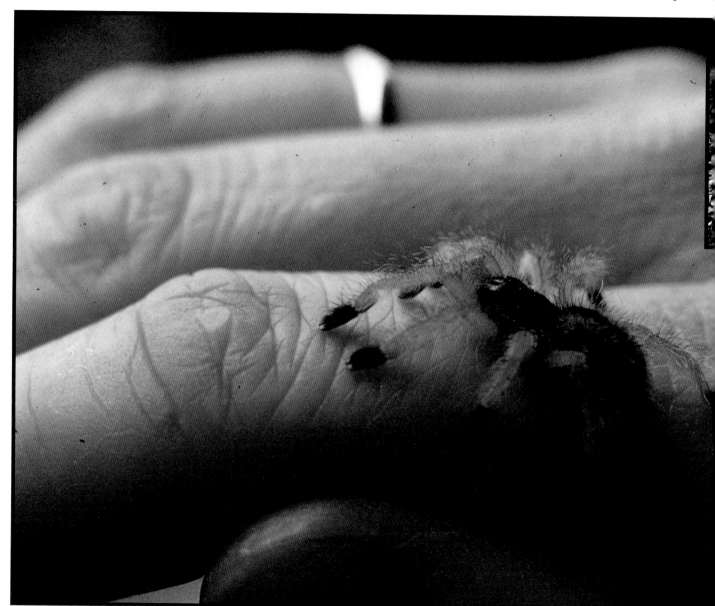

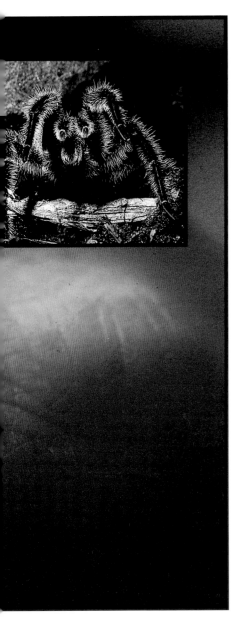

A Wolf spider carrying young on her back

Garden spider spiderlings emerging from their nest

Chapter 3
All the colours of the rainbow

Many of the most striking colours of the natural world are found among arthropods. Butterflies, ladybirds and caterpillars are familiar coloured arthropods, but there are many more exotic ones, such as gold bugs and red land crabs.

red – Eleven-spotted ladybirds

orange – Monarch butterfly

yellow – Longhorn beetle

een – Golden chafer bug

blue – Lycaenid butterfly

purple – Locust

How are these spectacular colours produced?

Arthropod colours are produced in three different ways.

Light is composed of a whole spectrum of colours. When it falls on an object some colours are absorbed and others reflected. This crab is red because pigments in its body absorb all the colours of the spectrum except red, which they reflect. This phenomenon is known as chemical colour.

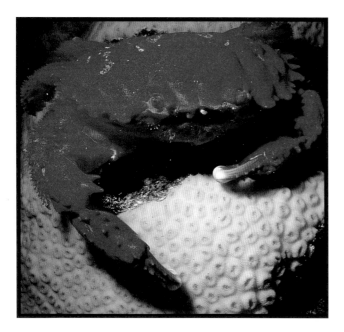

Many butterflies owe their iridescence to light playing on the scales of their wings. Although the scales themselves are unpigmented, they produce intense shimmering colour when light falls upon them. This is known as structural colour since it is caused by an interaction between light and thin layers that make up the scales. Which colours are reflected depends on the thickness of the layers and the gap between them, as well as the angle at which the light hits the layers. Because these vary over butterflies' wings, the wings appear multicoloured. The effect is the same as when light falls on an oily puddle and thin films of oil reflect colours.

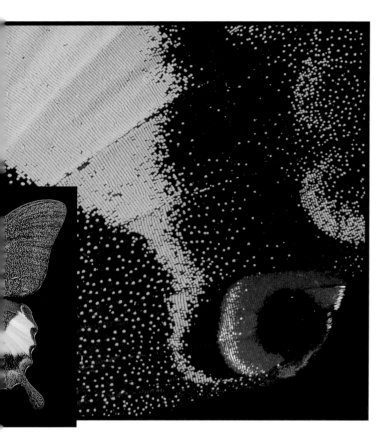

Glowing colours

Arthropods such as glow worms, do not need an outside source of light to glow. They glow as a result of the reactions of chemicals in their bodies, which produce light. This phenomenon is known as bioluminescence.

rd wing butterfly specimen, photographed from a tly different angle, shows the effect of structural colour.

Uses of colour

Bait
These blue lights photographed in a New Zealand cave are glowing fungus gnat larvae. Flying insects lured by their bright glow are snared in sticky threads trailing from the larvae. The larvae haul in the threads and devour their victims.

Warning and display

The startlingly vivid colours of some arthropods help to frighten off predators, or to attract attention away from vulnerable body parts.

A peacock butterfly flashes 'eyes' on its wings at predatory birds.

Clashing stripes of red, yellow and black warn predators that arthropods can retaliate by stinging or biting, or that they taste unpleasant.

Common wasp

Monarch caterpillar

Blind sequoia millipedes glow with a greenish light, warning predators of their bitter taste.

Jester bug

Sex

Arthropods use different types of colour as sexual signals. In many species, males and females are coloured differently. These differences help arthropods to distinguish possible mates. Sometimes the colours are visible to arthropods but not to us – we have to use ultraviolet light to see them.

Camouflage

Blending in with the surroundings can often help to deceive predators. Many arthropods have evolved effective and frequently bizarre forms of camouflage.

Small White butterfly under ultra-violet light

Small White butterfly under normal light

Glow worms (which are actually beetles) are probably the most famous glowing arthropods. The glow is used to attract mates. At night the wingless females turn their glowing abdomens up to the sky to attract males flying overhead. The males have a subdued glow, perhaps to warn off other males.

The peppered moth is a well-known example of the way colour has influenced the survival of a species. During the nineteenth century the numbers of dark brown peppered moths increased while the lighter variety became less common.

Sargassum crab

This happened because the dark moths were better adapted to life in the increasingly grimy environment of industrial England – they merged in with their dark surroundings.

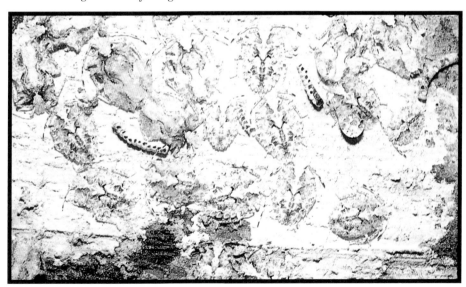

As drillers for sap, shieldbugs spend their time exposed on the surface of trees, vulnerable to attack from birds and other predators. Effective camouflage is their first line of defence.

Some species of caterpillar have shading on their bodies that makes them look flat so that predators, looking for solid prey, fail to pick them out from their background.

Several species of spiders live on the bark of trees; blending into their background is essential for them as they lie in wait for prey.

An Australian lichen spider has tufts of hairs on its legs which match lichen-covered bark.

Many jumping spiders hunt against a background of tree bark. They are often superbly camouflaged.

Jumping spider on the prowl

Mimicry

Other arthropods are even more skilful in the art of camouflage – rather than simply being the same colour as their background. They actually mimic some aspect of it.

Merveille du jour

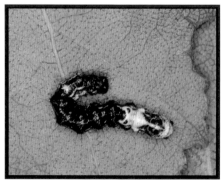

Some arthropods that live on trees or plants evade their predators by resembling fresh bird droppings!

A female praying mantis guards a brood of youngsters. The mother resembles the stem of the thorn bush on which she sits while her offspring mimic the actual thorns.

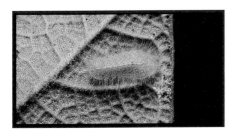

As leaves are common in terrestrial habitats it is not surprising that many insects have come to look like them.

While resembling a leaf in shape, the body of this bush cricket (from Trinidad) also incorporates features of a dead or dying leaf. The cricket even copies symptoms of fungal attack and the blotch mines caused by leaf-mining moths.

Resting on a leaf in a Malaysian rain forest, a bush cricket not only matches the colour of its surroundings perfectly but also includes leaf veins and disease blotches to enhance the deception.

Some arthropods use deception to evade their predators. Although they resemble dangerous arthropods, they themselves are harmless. Their warning colours prevent them from being eaten.

Many wasps are capable of inflicting a painful sting. By looking like a wasp, a harmless hover fly deters its predators.

Mutillid wasps have the most painful sting of all insects – not suprisingly they are frequently mimicked.

Some ground beetles are such good wasp mimics that scientists can identify precisely which species of mutillid the beetle is mimicking.

Many spiders also mimic mutillid wasps. This jumping spider resembles several species of Malaysian mutillids.

Its abdomen (on the left) looks like the wasp's head while the spider's head is on the right.

Chapter 4
Battling for survival

The range of arthropod weapons

In general, adult arthropods are well equipped for battle. A scorpion, for example, has jagged jaws, powerful pincers and a poisonous sting in addition to its thick, protective exoskeleton.

Claws and jaws

Many other arthropods have similarly effective weapons. Crabs and lobsters have pincers which, in a few species, are strong enough to remove a human finger.

European lobsters live in offshore waters. To escape predators they often retreat, tail first, into a crevice. Powerful pincers then confront the attacker.

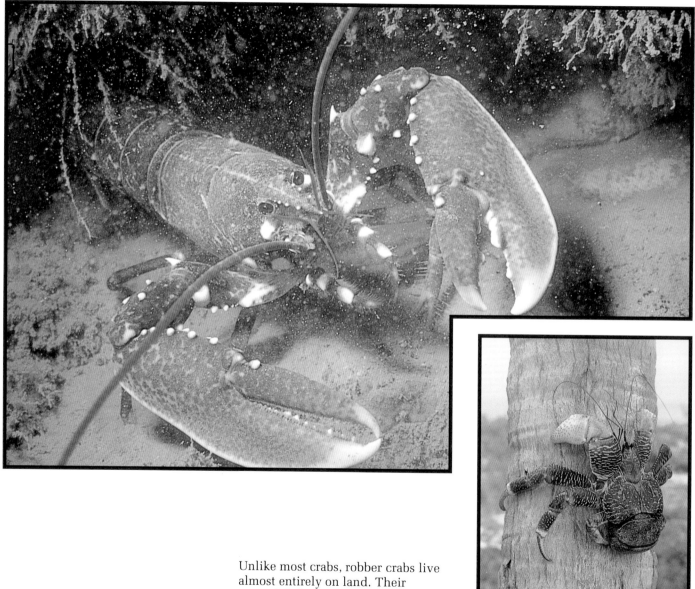

Unlike most crabs, robber crabs live almost entirely on land. Their powerful claws can break into the coconuts which they eat.

Poisons

Another very effective weapon in both attack and defence is a poisonous sting. Scorpions are well known for their stings but although all species are poisonous most are not deadly to humans.

Many spiders also use poison to subdue or kill prey but only a few species are deadly to humans. Probably the best known is the black widow spider.

Although this North American species is small, less than 3 centimetres across, the female's poisonous bite may be deadly if untreated.

Another extremely poisonous spider is the Sydney funnel web. Funnel web spiders belong to a group which includes bird-eating spiders or tarantulas. Most of this group are burrowers and the Sydney funnel web is no exception. Its burrow has a silken funnel at the entrance from which the spider gets its name.

Catching their prey

Most carnivorous arthropods have to hunt down or trap their food. To do this they have evolved highly ingenious methods of capture.

Sticky spider's webs are superb traps for flying insects.

Bolas spiders capture their victims with a sticky ball which they swing from a line. They then pull the ball, with prey attached, back up to the lair...

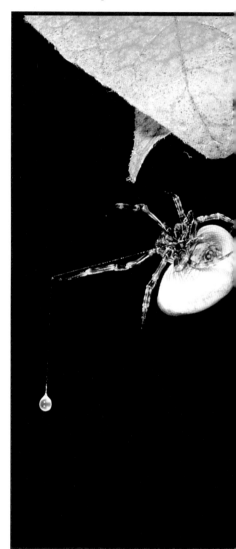

Trapdoor spiders rely on surprising their prey. The spider digs a burrow and covers it with a silken lid. Then, leaving the lid fractionally open, it lurks at the top of the burrow.

If a suitable animal strays near the lid the spider lunges out and bites it with its poisonous fangs. The prey is then dragged down into the burrow and consumed.

Dragonfly nymphs also take their prey by surprise. They conceal themselves in vegetation or mud on stream beds and lie in wait. When a small fish or tadpole approaches, the nymph shoots out its extendible jaws to grab the prey.

This strange creature is a caprellid, a marine crustacean which catches its prey in the same way as the praying mantis. It perches, unseen, on fronds of seaweed, waiting to grasp unsuspecting prey between its powerful pincers.

Virtually invisible against a background of plant leaves, the praying mantis waits for prey to wander within its reach. With its powerful forelegs folded in front as though praying, it is poised to pounce.

The mantis eats its prey alive.

47

Ant lion larvae (the immature stage of a dragonfly-like insect) ambush their prey. After digging a conical pit the ant lion larva lies buried at the bottom, with only its enormous jaws protruding, waiting for its victims.

When an ant scurries over the edge of the pit, the ant lion larva is alerted and flicks sand to knock the ant down into the bottom of the pit.

Then it seizes the ant between its large powerful jaws and begins to eat.

With their numerous long legs, centipedes are well adapted for high-speed hunting. Voracious carnivores, they feed on worms, snails and other arthropods, or even on lizards, toads and mice in the case of the

Tiger beetles have long legs for rapid pursuit and enormous biting jaws with which they can crush even the hardest outer skeleton. They will eat any animal small enough to fit between their jaws.

Speed and strength rather than stealth and cunning help some arthropods to catch their prey. Many have long legs and greatly enlarged, sometimes poisonous jaws which make them efficient hunters and killers.

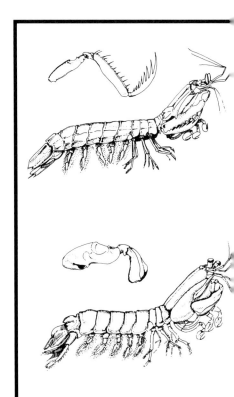

largest tropical centipedes. They kill their prey and paralyse with a bite from their poisonous fangs.

Food supplies for arthropod young

Many wasps lay their eggs on top of a living food source. Later their developing larvae will consume it.

One example of this is the spider wasp. The sting of this large wasp can paralyse a spider three times its size.

It then drags the spider into a hole which it has prepared and, after laying a single egg on top of the spider, covers the hole. When the wasp larva hatches it eats the spider live.

Mantis shrimps, like the land-living praying mantis, have powerful forelegs and use the element of surprise to catch their prey. They have evolved into two distinct types depending on the food they eat – some have spiny forelegs on which they spear small fishes, while others have powerful forelegs to smash the shells of prey such as snails and crabs.

The action of the 'spearer' when catching its prey is one of the fastest animal movements known – taking less than eight-thousandths of a second.

A 'smasher' can deliver a blow with a force matching that of a small calibre bullet – enough to break out of a glass tank.

Chapter 5
Food and feeding

How are arthropods able to thrive in so many different habitats? One reason for their success is that, as a group, they are able to exploit virtually any natural substance as a food source.

Arthropods devour a wide range of substances...

Devil's coach horse beetles eat other insects and slugs

Booklice eat moulds growing on paper

Tobacco beetles eat tobacco

Carpet beetles eat woollen carpets

The range of arthropod foods

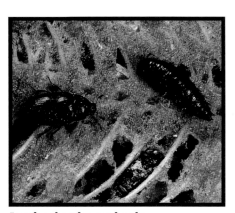

Leather beetles eat leather

. . . Some more attractive to humans than others!

Robber crabs eat coconuts

Locusts eat leaves

Lobsters eat fish

Mouthparts and limbs for the job

To consume this extraordinary range of foods, arthropods have evolved specially modified mouthparts and limbs. This may have happened to a much greater extent in arthropods than in other animals because their chitinous skeletons were more readily adaptable than internal bones.

House flies vomit digestive juices which reduce solid food to a sort of liquid soup. The liquefied food is then soaked up through the labellum - a sponge-like extension of the mouth.

Although mosquitoes are notorious blood suckers it is only the female that requires this diet to ensure the successful maturation of her eggs. She is attracted to a host animal's skin by its moistness and warmth. Once settled, the mosquito injects an anticoagulant into the animal's blood to prevent clotting and then sucks up the liquid meal through its syringe-like mouthparts.

Aphids feed on the sap of trees and plants, piercing the plant with sharp pointed mouthparts, called stylets. Pressure of the sap in the plant carries it up the stylets and into the animal's gut.

Butterflies suck up nectar through their long tubular mouthparts, which are kept coiled up like a hose-pipe when not in use.

Strainers

By opening its feathery limbs like parasols an atyid prawn can filter plankton from fresh water. The food is then transferred by these limbs to the mouthparts.

Barnacles strain sea water for food with their modified net-like limbs.

The large claw of this fiddler crab is not used for feeding but for signalling to females and for the defence of its territory.

To feed, the crab uses its small pincer to carry minute particles of mud to its mouth where anything edible is strained out.

Raspers

Horseshoe crabs, protected by a domed shield, burrow their way through the top layer of sand or mud on the sea-bed. They use jaw-like extensions on the base of their walking legs to trap prey such as clams and worms. They grind up their meal with these 'rasping' limbs before they pass it forward to the mouth on the underside of their bodies. The limbs of many simple arthropods, such as those of the horseshoe crab, are often used for both movement and feeding.

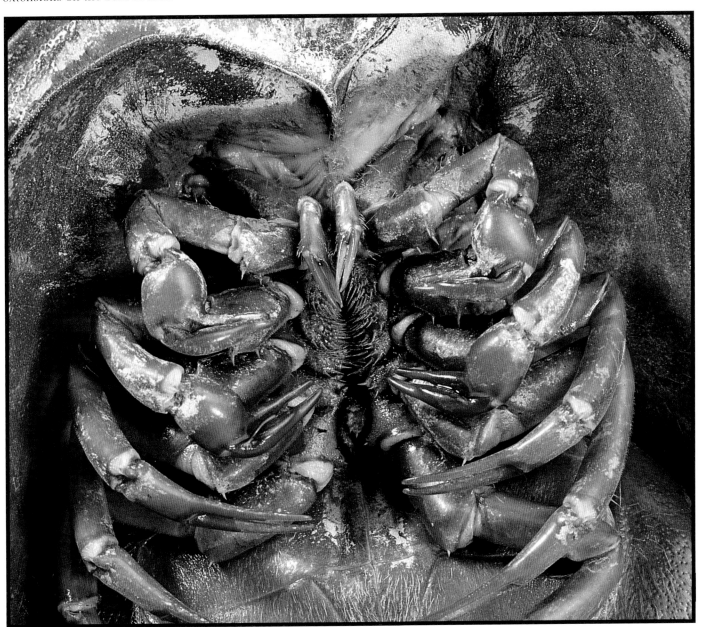

Chapter 6
Masterbuilders

The range of arthropod buildings

Many arthropods, particularly some insects, have evolved remarkably complex building skills and are able construct a wide variety of nests and shelters.

Garden spider

Leaf-cutter bee

Land crab

Mud-dauber wasp

Weaver ant

Leaf-rolling cricket
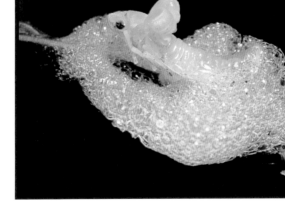
Froghopper

Leaf-rolling beetle

Sand wasp

Building techniques and uses

To build such a great variety of structures, arthropods employ a number of different materials and techniques depending on the purpose of the structure.

The building of a sticky silk net to catch prey, particularly prey that flies, is common amongst spiders but they are not the only arthropods to use silk nets.

The caddis fly larva, like many immature insects, is aquatic. It spins an elaborate funnel-shaped net and anchors it to aquatic plants or twigs. The running water of the stream in which it lives keeps the funnel open, and the net strains small organisms out of the water for the larva to eat at leisure.

Apart from building traps, arthropods also build shelters. Some are permanently fixed in one place and may be communal dwellings. In other cases an individual may build a portable home which it carries around. Most caddis larvae do this. They build a silk cocoon and then decorate it with dead plant matter or particles of sand and gravel. Some species attach shells and tiny snails to the outside of the cocoon to camouflage its outline. The exact method of construction and materials used vary from species to species.

In general, larval stages are more vulnerable than the adult stage, so most arthropod structures are concerned with nest building and protection of the eggs or young. Many arthropods such as bees, leaf-cutter ants and leaf-roller crickets, make use of plant material to build nests. Others, such as digger and mud dauber wasps make use of materials such as sand, mud and clay to construct their nests. Some simply excavate a hole in the ground while others construct elaborate fortresses. Many of the social wasps build amazingly intricate communal nests using a mixture of wood shavings and saliva. The variety of materials and techniques seems to be endless.

Inside a hive

A honey-bee comb is a marvel of engineering.

It consists of a series of carefully constructed hexagonal wax cells. If the cells were round or even octagonal or pentagonal there would be empty spaces between them. This would not only be a waste of space but would necessitate the building of separate walls for all or part of each cell. These difficulties can be avoided by using triangles, squares or hexagons – but of the three the hexagon has the smallest circumference for its area and is therefore the most economical to use.

Apart from making the most of the space available, the way in which hexagons fit together gives the comb great strength. A comb measuring 37 by 22 centimetres and containing only 40 grams of wax can hold almost 2000 grams of honey.

To begin building, bees gather together into a dense ball within which they maintain a temperature of 35° C – the temperature needed to produce wax. It is secreted from special glands in the bees' abdomens.

In a conventional hive, building begins at the top of a frame, usually at two or three different places at once. While the walls of the first cells are being constructed, new cells are started lower down the frame. Eventually they join to form the complete comb. The cell walls are built at a gradient of 13° to prevent honey from running out. Cell walls are only 0.073 millimetres thick with a margin of error of less than 0.0002 millimetres! Truly astounding precision.

Spiders' webs

Although all spiders can produce silk only some kinds spin it into webs. The silk they use (produced by special spinning glands in the spider's abdomen) is a remarkable material – it is twice as strong as steel, finer than human hair and can stretch to three times its own length. It is also edible, so before building a new web a spider may eat the old one and recycle virtually all the silk.

The typical spider's web, such as might be seen in a garden, is known as an orb web and consists of three major elements: frame threads which form the web's outline and serve as tethers; radial threads which converge into a central point; and the catching spiral. Only the catching spiral is sticky.

In the middle of the web is an area known as the hub. It consists of a concentration of strands although in some webs the hub is missing completely and the web appears with a large hole in the centre. Between the hub and the catching spiral is the 'free zone.' This enables the spider to change quickly from one side of the web to the other depending on which side the prey has been snared.

With more than 2000 species of orb-web spiders there is great variation in web architecture. By using a minimum of material (less than 0.5 milligrams of silk) and time (about an hour to build a typical web) spiders have an extremely efficient method of catching prey.

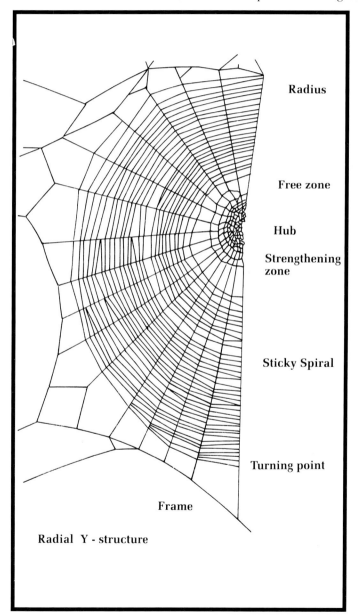

Radial Y - structure

Termite towers

Perhaps the most spectacular examples of arthropod structures are termite mounds. There are more than 2000 species of termites, most of which live in tropical or semi-tropical regions. Unlike most insects, termites cannot withstand too dry an atmosphere. Thriving in warmth and high humidity, they design complex nests to achieve these conditions.

Depending on the local climate, termite mounds come in a variety of shapes.

Some termites live in areas of very high rainfall, such as tropical forests. To protect themselves from the torrents, they build extraordinary roofs on their towers – making the structures look like a series of mushrooms stacked one on top of the other.

Compass termites live in the Australian outback. Their towers are up to five metres long and three metres high but very narrow. Their short sides face exactly north-south so that the surface exposed to the rays of the midday sun is small, while the long sides catch the morning and evening sun. In the cold season, termites congregate on the east side in the morning and the west side in the evening in order to gain maximum heat from the sun.

The interior architecture of many termite mounds is even more impressive than the exterior. Chambers are laid out according to their different functions which may indicate a definite building plan. Inside some mounds there may be as many as two million individuals. Their oxygen consumption is considerable and without ventilation they would all suffocate within a day. It is only by an ingenious system of air conditioning that they avoid this fate.

The nest, which may be two metres across, sits like a giant lollipop on a stick. Above it rises an empty hollow tower up to three metres high. When the nest gets too hot, worker termites rush to open a valve made of dried mud at the top. Hot, stale air escapes from the nest and rises up the tower. Near the top, the walls of the tower are so thin that the stale air, rich in carbon dioxide, seeps out while fresh air floods in. This cooler air sinks to the bottom of the tower and enters the nest from below. By opening and closing the valve in their nest the termites have complete control over the air flow through their underground environment.

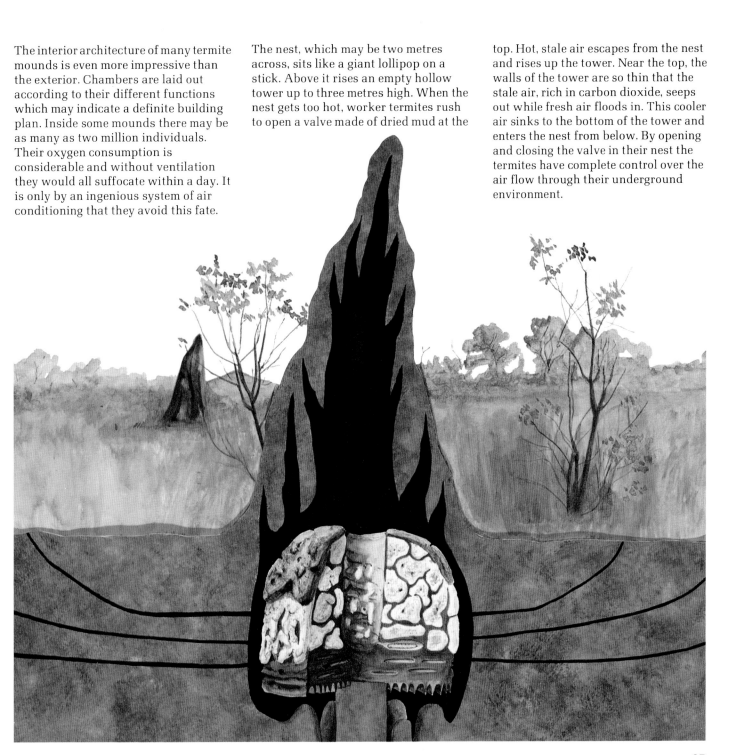

Chapter 7
Taking up residence

Many arthropods are familiar to us because we come across them every day in our homes. Many kitchens contain thriving arthropod communities – beetles inhabit the woodwork, silverfish and woodlice live under the sink, ants forage the food cupboards, cockroaches lurk behind the cooker and bluebottles buzz around the food.

Bedrooms have their own arthropod residents – people who sneeze when they get into bed should blame the dust mites that gather in the mattress. In some houses bedbugs swarm behind the wallpaper, coming out at night to feed on sleeping humans. Many other arthropods can be found throughout the home – perhaps the one we know best is the house spider, which patrols our rooms in search of a mate.

Bluebottle

Dust mite

Silverfish

Bed bug

Why do these unwelcome visitors invade our homes?

For carnivorous arthropods, such as the house spider and the Devil's coach horse beetle, our homes are an easy source of prey.

Some arthropods like to feed on our food. For bluebottles and houseflies, our wastebins are a gastronomic delight – rotting, oozing food attracts them because it is easy to turn into a solution which they can mop up with their sponge-like mouthparts.

Garden ants troop into the house in search of food which they can take back to their nests.

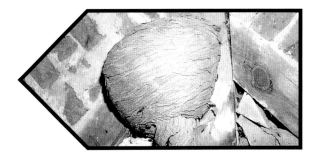

Wasps build nests in the safety of the attic. This nest was started by a single queen. Now she lives there with 5000 of her offspring.

Clothes moths seek out our woollen fabrics to lay their eggs in. When the larvae hatch out, they munch their way through our clothes.

Parasitic arthropods, such as fleas, ticks and lice, are carried into our homes by us or our pets. Fleas, from our dogs and cats, lay a large number of eggs. Their larvae feed on detritus, especially flea droppings, and can reach maturity in less than a month.

Some arthropods are even specialized consumers of our household furnishings. For instance, the larvae of the carpet beetle slowly consume our woollen carpets; booklice feed on moulds growing on paper and the glue holding our books together; and woodworms, the larvae of furniture beetles, tunnel around in our woodwork as they develop.

Carpet beetle

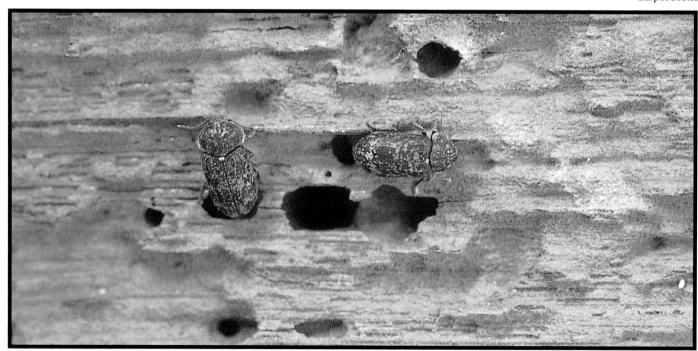

Furniture beetle

An arthropod in every bite!

Many arthropods find a congenial home amongst our food stores.

Larder beetles creep into our cupboards to find a suitable place to lay their eggs.

The newly hatched grubs gorge themselves on our supplies of bacon.

Cheese mites lay their eggs on cheese - this mature cheddar is infested with them

Spider beetles and their larvae are major pests of stored food. Apart from eating spices and sauce mixes, they feed on textiles and other apparently indigestible things.

Biscuit beetles infest our biscuits, but, fortunately for chocolate lovers, tend to choose the dry crispbread variety rather than chocolate-coated ones.

Open bags of flour can quickly become teeming with mites, caterpillars and beetle grubs.

Infestation – how can we stop it?

Some arthropods have benefited from modern improvements in housing, but many others have found them disastrous. Central heating, wall to wall carpeting and insulation have made our houses havens for heat seeking arthropods, such as pharaoh ants, but they have also threatened the damp habitats of woodlice, silverfish and spiders. The tiny clover mite, which moults and lays eggs in crevices in external brickwork may wander into our homes but be thwarted by the high quality finish on our walls. The once common bedbug is now rare in most developed countries.

Domestic pests have not all disappeared. Ants still invade our food cupboards, wasps and flies still buzz around our kitchens, and spiders still scurry across our floors.

Grain weevils

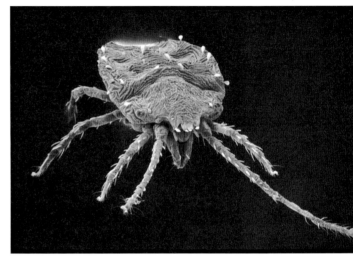
Clover mite

Woodlice

Silverfish

To prevent these pests from breeding in or near our homes our dustbins should be kept covered. We should also watch out for ants' or wasps' nests near our houses. Often these are easy to find because there are a suspiciously large number of insects flying or crawling around near them.

Inside the house, food should be kept under cover – preferably in the fridge or in cupboards which can be tightly closed and which are well above floor level.

There are many insecticides available for killing insects and spiders. But these products are unpleasant to use and although they may get rid of a few individual pests, they do not strike at the heart of an infestation. A better way to deal with pests is to keep them out in the first place.

Once inside the house the bathroom is one of the most common sites for an encounter with a spider. How many times have you found a spider scrabbling around in your bath? Contrary to popular belief, these spiders do not climb up the drainpipe and through the plug hole only to find themselves trapped. In fact, they usually fall down the slippery sides of the bath. Recently, a device for rescuing spiders in this situation has come on the market. It consists of a small rope ladder which can be dangled down into the bottom of the bath, giving the spiders an easy escape route. It is worth being kind to house spiders because they are completely harmless and they help to keep our homes free of more serious pests.

Chapter 8
Arthropods and humans

Disease carriers

Some arthropods pose a major threat to human health, especially in the hotter parts of the world. Many tropical insects, mites and ticks feed on human blood and this is one of the commonest ways in which some serious diseases are spread. Arthropods are particularly prevalent in the tropics so it is here that many of the most debilitating human diseases occur.

Anopheline mosquito feeding

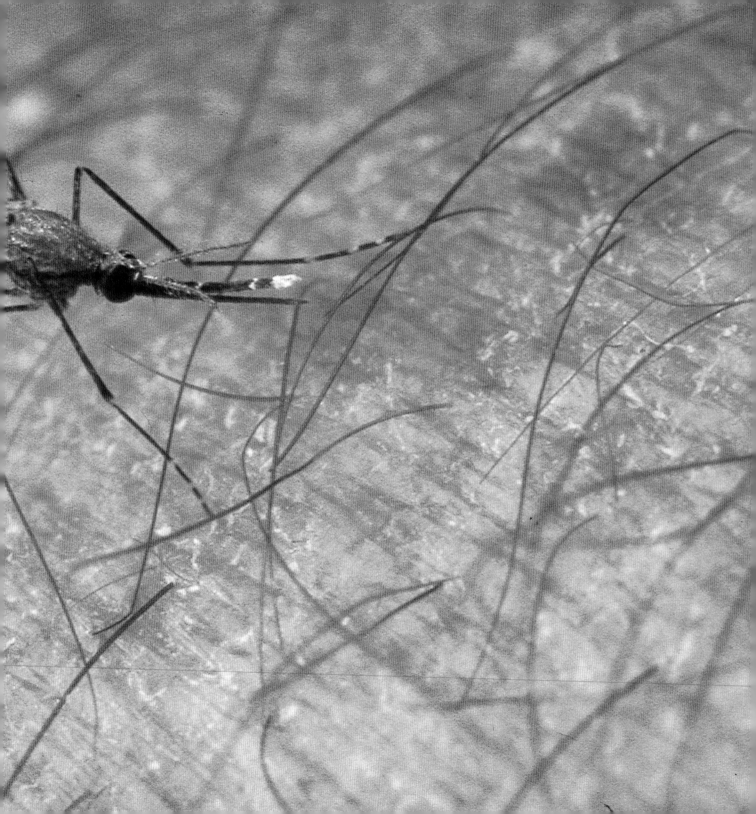

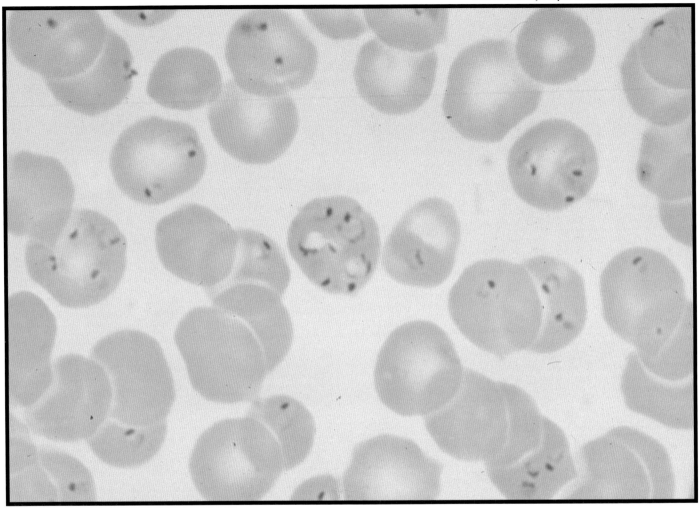

Plasmodium falciparum in human blood cells

Arthropods themselves are usually not the direct cause of diseases but, by feeding on man and other animals, they spread the microscopic organisms which produce the symptoms.

Arthropods which transmit diseases are known as vectors and different diseases have different vectors. For example, anopheline mosquitoes are the vectors for malaria and filariasis; blackflies for river blindness; tsetse flies for sleeping sickness; body lice for typhus; fleas for plague; and certain kinds of ticks for Lyme disease and some types of relapsing fever.

Of all the diseases carried by arthropod vectors, malaria is probably the most widespread. It has been estimated that, even today, there are at least two billion people at risk from malaria.

The main vector of malaria in Africa is the mosquito *Anopheles gambiae*.

Male mosquitoes do not feed on blood so only females carry the disease. To pass on the disease a mosquito must be infected with the malarial parasite, a tiny single-celled organism called *Plasmodium*.

When an infected female *Anopheles* mosquito feeds on human blood, these organisms present in the mosquito's saliva, invariably get passed into the bloodstream. Their presence causes a raging fever – the sufferer alternates between feeling very hot and very cold with the fever returning every few days if not treated. In the worst forms of malaria sufferers may die.

River blindness today affects about 20 million people in Africa and central South America. The causative organism is a microscopic worm but the arthropod vectors responsible for transmission are various species of blackflies.

Since blackflies breed in fast-flowing water, rivers and irrigation ditches are favourite sites for infestation. By feeding on human blood, blackflies pass the worms into the bloodstream. They breed and spread to the skin and eyes. If untreated the disease can impair vision leading ultimately to total blindness.

African victims of river blindness being led with sticks

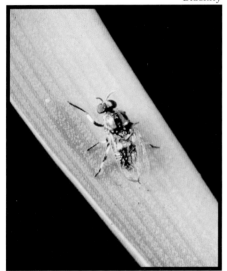

Blackfly

Another fly, the tsetse, carries a disease known as sleeping sickness. Like malaria and river blindness, the causative organism is microscopic and is passed to humans in the saliva of the fly when it takes a blood meal.

Sufferers from sleeping sickness become increasingly weak and tired as the disease takes hold. They eventually become too weak to eat and die.

Sleeping sickness currently affects 35 African countries and puts an estimated 36 million people at risk. However, the disease also affects cattle and about a quarter of Africa, unable to sustain cattle for this reason, cannot be used to its full potential.

Tsetse fly feeding

Crop destroyers

Although arthropod vectors carry many serious diseases it is not always humans and livestock that are under attack – arthropods can also be serious crop destroyers.

More than one-third of the world's crops are lost to arthropod pests. In Africa and Asia almost 50 per cent of crops are lost in this way. Of all the arthropods, insects are the most voracious consumers of food crops.

Locusts are among the best known and most spectacular insect pests. Occurring mostly in Africa, these insects can congregate into massive swarms of many millions of individuals. These swarms may weigh hundreds of thousands of tonnes. Such a swarm can strip vast areas of greenery in a very short time.

Although few arthropod pests occur in such large numbers, many others also cause devastation. Here are some of the worst:

Brown planthopper – a major pest of rice in Asia

Nettle caterpillar – a major pest of coconut and oil palm in South East Asia

Aphid – a sap-sucking pest

Colorado beetle – the world's worst pest of potatoes

Armyworm – Africa's most notorious cereal pest

In fact for *any* given crop there are dozens of arthropod pests – the problem remains how to control them.

Fighting back

Any attempt at effective control of arthropods, whether as disease carriers or crop destroyers, must take into account several factors. Most important is a detailed knowledge of the biology and ecology of the arthropod concerned, the animal or plant affected and, if relevant, the disease-causing organism involved.

Efforts to control vector-borne diseases have, not surprisingly, centred on control of the vector itself because without these intermediaries, diseases would not spread.

In the case of malaria, the insecticide DDT was used with some success in the 1940s, effectively eradicating malaria from southern USA and Europe. Parts of Africa and South America were also cleared of the disease. However, by 1950, two species of mosquito had already become immune to DDT. This, coupled with lack of funds, poor management and general complacency about the disease led to a resurgence of cases in areas formerly free of malaria. Nowadays more simple preventative methods which remove mosquito breeding sites are often preferred to nonspecific use of insecticide.

Modern pest control methods often try to combine cheapness with efficiency and environmental safety. Some success has been achieved in Zimbabwe using a very simple trap for tsetse flies.

Tsetse flies often feed on cows and are drawn to their shape and smell. Scientists researching tsetse fly behaviour discovered that, after a cow shape, tsetse flies are most attracted to a square – especially a black one. Using this information some simple black, square traps were constructed with insecticide-impregnated netting attached. Scientists even managed to make the traps smell like cows. When the flies approached the trap they flew into the fine netting and were killed by the insecticide.

Unfortunately, not all pests can be controlled using simple traps. Sometimes an arthropod's lifecycle is complex and the best way to stop the pest unclear. Most modern pest control methods used in the field rely on a combination of the best available techniques tailored for the particular pest concerned. Preventative measures which educate the local population are often the easiest way to begin fighting back.

Chapter 9
Ancient arthropods

The astonishing versatility of arthropods has contributed to the evolution of the huge variety of forms we find today. Like all living things, today's arthropods have evolved from earlier forms – those that could adapt to changes in their environment survived, while those that could not became extinct.

What do we know about these early arthropod forms and what can they tell us about the evolution of arthropods?

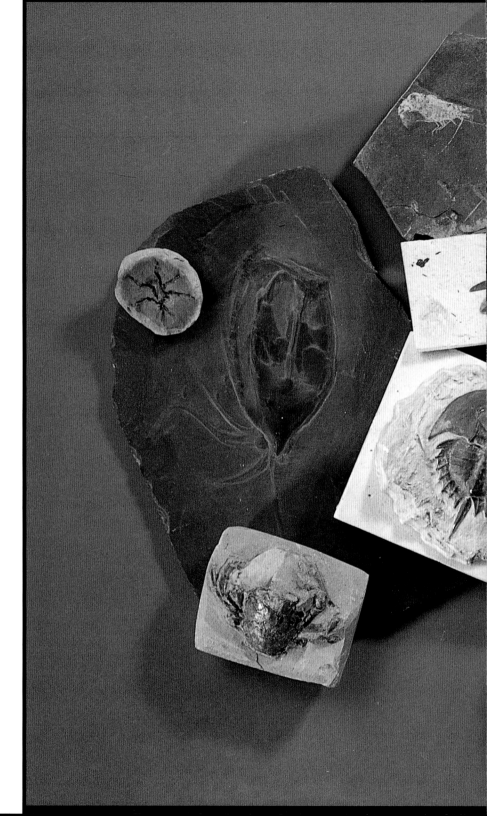

Bringing fossil arthropods back to life

We know that arthropods have been around for millions of years because portions of their bodies or traces of their activities, such as trails and burrows, have been found preserved in rocks up to 600 million years old. Apart from some worm-like animals, few other creatures are known to have existed at this time.

Some of these early arthropods were very different from those around today.

... This giant water scorpion was one of the largest arthropods that ever lived. The group died out 250 million years ago.

These old fossils, often distorted or broken, are the only evidence we have of such ancient creatures. But, using them, we can sometimes go back in time to reconstruct the lives of the early arthropods – to find out what they ate, how they moved, where they lived, and so on.

Judging by the fossils, some ancient arthropods would have looked almost identical to modern arthropods. Because their similar body parts were probably adapted to perform the same functions, it is likely that these early arthropods lived in much the same way as their modern relatives.

Ancient horseshoe crabs, and modern horseshoe crabs living 150 million years later, would probably be indistinguishable. Like their modern relatives, the first horseshoe crabs probably lived deep in the sea, emerging only in the spring at full moon and high tide to spawn on beaches.

Fossils are harder to interpret when there are no modern equivalents with which to compare them. In such cases it often helps to look at several fossils for clues as to how an animal lived.

This is the fossil of a trilobite, an ancient sea arthropod. The fossil shows its flattened body.

A second fossil shows a trilobite rolling up.

A third fossil shows a trilobite rolled up tighter.

A fourth fossil shows a trilobite in a tight ball.

Maybe trilobites rolled up into a ball when threatened — just as the pillbug millipede does today.

The incomplete history of arthropods

Collections of fossils of the same age may help us to investigate the lifestyles of the early arthropods, but they cannot explain how these creatures originated or what happened to them subsequently.

Which of the strange forms below were the ancestors of modern arthropods and which of them vanished, having failed to adapt to a changing Earth?

To answer these questions we would need to look at a series of fossils of different ages. Only then could we see how a particular arthropod had evolved over millions of years. In practice, however, this is rarely possible as most ancient arthropods vanished without trace.

Fossil survivors

The extent to which an animal is fossilized depends on where it dies and how tough its body parts are.

The best preservation occurs in mud, which, under compression, eventually hardens to stone. So, usually, fossilized arthropods are ones that lived in or near water.

The great majority of fossils are of hard parts of an animal. So although many creatures may have been entombed in muddy sediment, few became fossils because their body parts were too soft.

Large arthropods are preserved better than small ones...

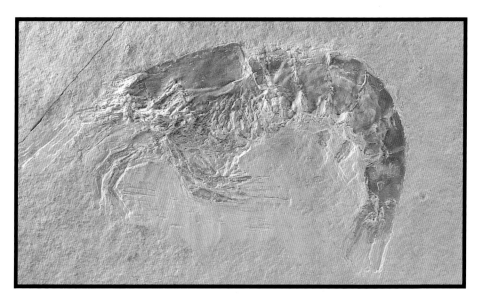

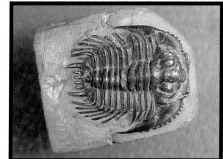

...and fossils of the more durable body parts – such as wings – are often found without the rest of the body.

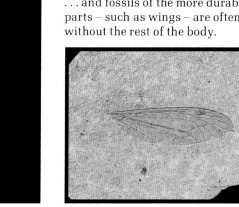

In some exceptional cases, it is possible to find fossils where even the softest parts are intact.

Fossil insects and spiders have been found exquisitely preserved in lumps of amber – the hardened resin of trees.

The Burgess Shale in Canada has been found to contain many fossils including arthropods which were exceptionally well preserved because they were caught in underwater mud slides and buried. Even the soft parts of animals which would normally have decayed and disappeared without trace have left impressions in the fine-grained shale.

Apart from giving us a glimpse of an ancient community, the Burgess Shale discovery has also helped to confirm the fact that many animals are *not* normally preserved as fossils.

Some of the animals found in the Burgess Shale are completely different from any living organisms, and may be the fossils of creatures that died out millions of years ago. Others may have been the precursors of modern forms, but without the discovery of intermediate stages we will never know for sure.

By 400 million years ago arthropods had left the sea and evolved into land-living creatures. Scorpions, mites and springtails – all adapted to surviving in damp areas – were some of the first land dwellers.

Great events in arthropod evolution

Six hundred million years ago, the first arthropods, believed to have evolved from a worm-like ancestor, lived in the sea. Today they are one of the most numerous and diverse of animal groups and are found in almost every part of the world.

What happened in between?

The fossil record, although incomplete, reveals some fascinating landmarks...

Insects, the first creatures to 'conquer' the air, appeared around 300 million years ago. Curiously, there are no fossils which show how wings developed – all the early fossils of insects have either well-developed wings or none at all.

Trilobites were once the most common kind of arthropod. At their most abundant, about 500 million years ago, they had spread through the seas and diversified into a great number of species. Then, 250 million years later, the group dwindled and died out. No one knows why.

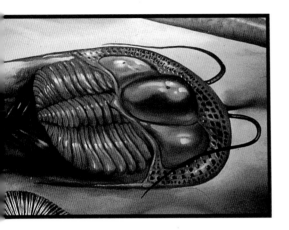

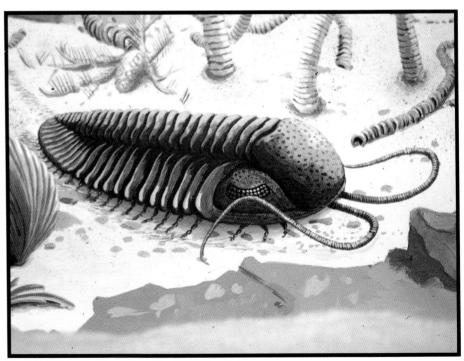

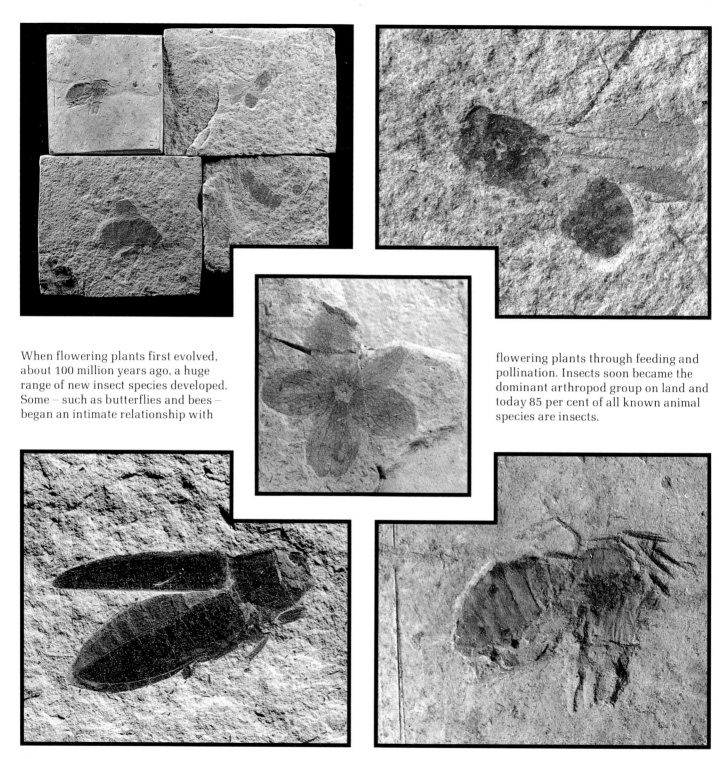

When flowering plants first evolved, about 100 million years ago, a huge range of new insect species developed. Some – such as butterflies and bees – began an intimate relationship with flowering plants through feeding and pollination. Insects soon became the dominant arthropod group on land and today 85 per cent of all known animal species are insects.

A last word

The modern arthropods featured in this book fall into four main groups:

Insects, which have three pairs of legs, wings, and one pair of antennae.

Arachnids, which have four pairs of legs and no antennae.

Crustaceans, which have two pairs of antennae (and almost all live in water).

Myriapods, which have bodies made up of many segments, each with one or two pairs of legs.

But there are some creatures that are difficult to place...

Velvet worms are soft animals with a caterpillar-like body. At first sight they seem unlikely to be related to arthropods, but the ancestors of insects and myriapods were probably similar.

'All-legs-and-no-body' sea spiders have four or more pairs of long jointed legs. Although they have a very different shape to arachnids, they may be related to them.

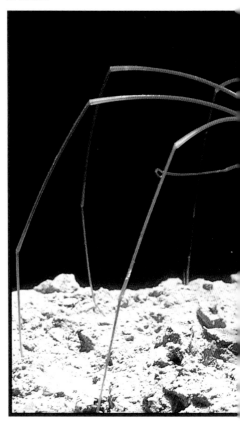

Water bears are tiny animals with four pairs of unjointed legs and a thin outer skeleton. Some scientists believe they are related to arachnids.

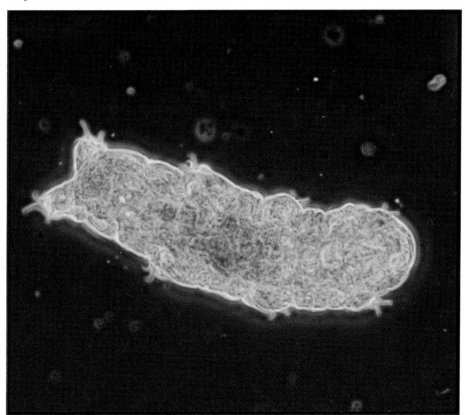

Because these creatures are alive today we can investigate them in much greater detail than fossils – not only can we study their general anatomy, but we can also probe into their cells to examine the genes that determine all their features.

This is one of the techniques scientists use to work out how different creatures are related – those with similar genes are more closely related than those with very different ones.

Such work can provide us with valuable clues about arthropod evolution. Every year some 12 000 new arthropod species are discovered. By studying these and continuing the search for new fossil arthropods, eventually we may be able to see how the 'primitive' ancient arthropods evolved into the most successful animal group on Earth.

Further reading

Chapter 1
General

Spiders of the World by R & K Preston-Mafham. Blandford Press Ltd 1984.

The Insects – structure and function 3rd ed. by R F Chapman, Hodder & Stoughton 1982.

How to Know the Mites and Ticks by B McDaniel, W M C Brown Co. 1979.

Arachnida 2nd ed. by T H Savory, AP 1977.

Insect Life by M Tweedie, Collins 1977.

Spiders, Scorpions, Centipedes and Mites by J L Cloudsley-Thompson, Pergamon 1968.

Crustaceans by W L Schmitt, The University of Michigan Press 1965.

A Biology of Crustacea by J Green, H F & G Witherby Ltd 1961.

Chapter 2
How arthropods grow

The Principles of Insect Physiology 7th ed. by V B Wigglesworth, Chapman & Hall 1972.

Introductory Insect Physiology by R L Patton, W B Saunders & Co. 1963.

Chapter 3
All the colours of the rainbow

Bioluminescence by E N Harvey, Academic Press 1952.

Chapter 4
Battling for survival

Defence Mechanisms in Social Insects by H R Hermann (ed.), Praeger 1984.

Biology of spiders by R F Foelix, HUP 1982.

Chapter 5
Food and feeding

Food and Feeding by G F Warner in *The Biology of Crabs*, Elek Science London 1977.

Chapter 6
The masterbuilders

Spiders webs, Behavior and Evolution by W A Shear (ed.), SUP 1986.

Animal Architecture & Building Behaviour by M H Hansell, Longman 1984.

Animal Architecture by K von Frisch & O von Frisch, Hutchinson & Co 1975.

Master Builders of the Animal World by D Hancocks, Hugh Evelyn 1973.

Chapter 7
Taking up residence

Common Insect Pests of Stored Food Products L A Mound (ed.) BM (NH) 1989.

Household Insect Pests by N E Hicken, ABP Ltd 1974.

Chapter 8
Arthropods and humans

Plant Pests and Their Control by P G Fenmore, Butterworths 1984.

Arthropods and Human Skin by J O'Donel Alexander, Springer Verlag 1984.

The Rise and Fall of Malaria in Europe by L J Bruce-Chwatt & J de Zulueta, OUP 1980.

Insects and History by J L Cloudsley-Thompson, Weidenfield & Nicholson 1976.

Arthropod Vectors of Disease by J R Busvine, Edward Arnold 1975.

Insects and Other Arthropods of Medical Importance by K G V Smith, BM(NH) 1973.

Chapter 9
Ancient arthropods

The Burgess Shale by H B Whittington, YUP 1986.

Fossils. The Key to the Past. by R Fortey, Heinemann 1982.

Arthropod Phylogeny by A P Gupta (ed.), Van Nostrand Reinhold Company 1979.

The Arthropoda by S M Manton, Clarendon Press 1977.

Trilobites by R Levi-Setti, UCP 1975.

Evolution in the Arachnida by T H Savory, Merrow Publishing Company Ltd 1971.

INDEX OF SCIENTIFIC NAMES AND ACKNOWLEDGEMENTS

Because of the nature of this book the index is set out in numerical order and reads across a double page spread, left to right and top to bottom.

Page no. Common name (*Scientific name*) Photographer/Agency

Chapter 1

8–9   Giant dragonfly *Meganeura* sp (NHMPU)

10–11   Water flea *Daphnia* sp (NHMPU)
Common lobster *Homarus gammarus* (*L. Pitkin*, NHM)
Red coral shrimp *Rhynchocinetes rigens* (*L. Pitkin*, NHM)
Velvet swimming crab *Necora puber* (*L. Pitkin*, NHM)
Woodlouse *Porcellio* sp (NHMPU)
Edible crab *Cancer pagurus* (*L. Pitkin*, NHM)

12–13   Giant centipede *Scolopendra gigas* (OSF)
Stone centipede *Lithobius* sp (NHMPU)
Soil mites *Pergamasus* sp (NHMPU)
Harvestman *Gagrella* sp (OSF)
Woodlouse spider *Dysdera crocata* (NHMPU)
Desert millipede *Orthoporus ornatus* (*J. Cooke*, OSF)
Desert scorpion *Androctonus australis* (*C. Betts*, NHM)
Stone centipede *Lithobius* sp (NHMPU)
Imperial scorpion *Pandinus imperator* (*S. Moore*, NHM)
Lace-web spider *Amanrobius* sp (NHMPU)

14–15   Desert locust *Schistocerca gregaria* (NHMPU)
Longhorn beetle larva *Stromatium barbatium* (NHMPU)
Holly blue butterfly *Celastrina argiolus* (NHMPU)
Burying beetle *Nicrophorus humator* (NHMPU)
Lesser stag beetle *Dorcus parallelipipedus* (NHMPU)
Praying mantis *Hierodula* sp (*J. Martin*, NHM)
Cockchafer *Melolontha melolontha* (NHMPU)
Peacock caterpillars *Nymphalis io* (*J. Goode*, NHM)
Common earwig *Forficula auricularia* (NHMPU)
Lacewing *Osmylus fulvicephlus* (NHMPU)
Common wasp *Vespula vulgaris* (NHMPU)
Blue mormon butterfly *Papilio polymnestor* (*J. Goode*, NHM)

Chapter 2

16–17   Giant spider crab *Macrocheria kaempferi* (NHMPU)

18–19   Shore crab *Carcinus maenas* (NHMPU)

20–21   Mayfly (larva) *Ephemera danica* (NHMPU)
Mayfly (adult) *Ephemera danica* (*S. Dalton*, NHPA)
Nymphalid butterfly (life cycle) *Lexias aeropa* (*J. Martin*, NHM)

22   Coxcomb prominent caterpillar *Ptilodon capucina* (*M. Tweedie*)

23   Edible crab (adult) *Cancer pagurus* (*J. & M. Bain*, NHPA)
Edible crab (larva) *Cancer pagurus* (*R. Ingle*, NHM)

24–25   Barnacle (larva) *Semibalanus balanoides* (*G. Boxshall*, NHM)

Barnacle (adult) *Chthamalus montagui* (*M. Tweedie*)
Spirostrepid millipede (*P. Hillyard*, NHM)
Crayfish (adult) *Astacus leptodactylus* (*L. & T. Bomford*, Ardea)
Crayfish (juvenile) *Astacus leptodactylus* (*J. Burton*, BC)
Giant centipede *Scolopendra canidens* (*P. Hillyard*, NHM)
Centipede moulting (*M. Tweedie*)
Cave centipede *Scutigera coleoptrata* (*M. Tweedie*)

26–27 Pink-toed tarantula *Avicularia avicularia* (*D. Thompson*, OSF)
Pink-toed tarantula *Avicularia avicularia* (*P. Sharp*, OSF) INSET
Wolf spider *Trochosa terricola* (*M. Tweedie*)
Garden spider spiderlings *Araneus diadematus* (*C. Jermy*, NHM)

Chapter 3

28–29 Eleven-spot ladybirds *Coccinella undecimpunctata* (*M. Tweedie*)
Monarch butterfly *Danaus plexippus* (NHMPU)
Longhorn beetle *Tragocephala nobilis* (NHMPU)
Golden chafer beetle *Anomala* sp (*J.H. Carmichael*, NHPA)
Lycaenid butterfly *Arhopala araxas philtron* (*J. Martin*, NHM)
Locust *Ornithacris pictula magnifica* (NHMPU)

30–31 Xanthid crab *Etisus* sp (*L. Pitkin*, NHM)
Swallowtail butterfly *Papilio kama* (NHMPU) TOP
Birdwing butterfly *Troides prattorum* (NHMPU)
Glow worms *Lampyris noctiluca* (*K. Taylor*, BC)

32 Fungus gnats *Arachnocampa luminosa* (NHPA)

33 Peacock butterfly *Nymphalis io* (*G.A. Maclean*, OSF)
Common wasp *Vespula vulgaris* (NHMPU)
Monarch caterpillar *Danaus plexippus* (NHMPU)
Sequoia millipede *Luminodesmus sequoiae* (*R. Hoffman*, RUV)
Jester bug *Graphosoma italicum* (*M. Tweedie*)

34–35 Small white butterflies *Pieris rapae* (NHMPU)
Sargassum crab *Portunus sayi* (*J.H. Carmichael*, NHPA)
Peppered moth *Biston betularia* (NHM)
Glow worm (male) *Lampyris noctiluca* (*M. Tweedie*)
Glow worm (larva) (*M. Tweedie*)
Shieldbugs *Coriplatus depressus* (*K.G. Preston-Mafham*, PP)

36–37 Eyed hawkmoth caterpillars *Smerinthus ocellata* (*M. Tweedie*) 36 T & B
Lichen spider *Pandercetes* sp (*R.&D. Keller*, NHPA)
Jumping spider *Sparteus* sp (*K.G. Preston-Mafham*, PP)
Jumping spider *Sitticus* sp (*J. Cooke*, OSF)

38–39 Merveille du Jour moth *Dichonia aprilina* (*M. Tweedie*)
Praying mantis *Oxyophthalmellus somalicus* (*K.G. Preston-Mafham*, PP)
Bird-dropping spider *Archemorus* sp (*M. Tweedie*)
Alder moth caterpillar (early) *Acronicta alni* (*M. Tweedie*)
Alder moth caterpillar (late) *Acronicta alni* (*M. Tweedie*)
Lycaenid caterpillar (*J. Martin*, NHM)
Trinidad bush cricket *Pycnopalpa bicordata* (*J. Cooke*, OSF)
Malaysian bush cricket *Chloracris* sp (*M. Tweedie*)

40–41 Robber flies *Laphria thoracica* (*K. Lewis*, OSF)
Mutillid wasp *Dolichomutilla sycorax* (*K.G. Preston-Mafham*, PP)
Hoverfly *Sericomya silentis* (*M. Tweedie*)
Ground beetle *Eccoptoptera cupricollis* (*K.G. Preston-Mafham*, PP)
Jumping spider *Orsima* sp (*K.G. Preston-Mafham*, PP)

Chapter 4

42–43	Desert scorpion *Scorpio* sp (*C. Bett*, NHM)
44–45	Common lobster *Homarus gammarus* (*L. Pitkin*, NHM) Black widow spider *Lactrodectus mactans* (*K.G. Preston-Mafham*, PP) Robber crab *Birgus latro* (*S. Blackmore*, NHM) Sydney funnel-web spider *Atrax robustus* (*R.W.G. Jenkins*, NHPA)
46–47	Hoverfly *Eristalis tenax* (*S. Dalton*, NHPA) Bolas spider *Mastophora bissacata* (*G. Murrey* Jnr., OSF) Dragonfly larva *Aeshna* sp (NHMPU) Caprellid *Caprella* sp (*P. Parks*, OSF) Trapdoor spider *Ummidia* sp (*J. Cooke*, OSF) Praying mantis *Mantis religiosa* (*Z. Leszczynski*, OSF)
48–49	Ant lion larva *Formacaleon tetragrammicus* (*P. Ward*, BC) Giant centipede *Scolopendra gigas* (OSF) Tiger beetle *Ciccindela sexgutata* (*J. Shaw*, NHPA) Spider wasp *Hemipepsis indica* (*A. Bannister*, NHPA)

Chapter 5

50–51	Desert locust *Schistocerca gregaria* (*S. Dalton*, NHPA)
52–53	Devil's coach horse beetle *Staphylinus olens* (*M. Tweedie*) Booklouse *Liposcelis bostrychophilus* (NHMPU) Tobacco beetle *Lasioderma serricorne* (*N. Cattlin*, HS) Robber crab *Birgus latro* (*G. Claridge*, NHPA) Desert locust *Schistocerca gregaria* (*S. Dalton*, NHPA) Carpet beetle *Anthrena verbaski* (OSF) Lobster *Homarus americanus* (*Z. Leszczynski*, OSF) Leather beetle *Dermestes* sp (*G.I. Bernard*, OSF)
54	Bluebottle *Calliphora vomitoria* (*G.I. Bernard*, OSF)
55	Anopheline mosquito *Anopheles* sp (NHMPU) Aphid *Amphorophoro tuberculata* (NHMPU) Butterfly *Opsiphanes* sp (*J. Cooke*, OSF)
56–57	Atyid prawn *Atya* sp (NHM) Acorn barnacle *Semibalanus balanoides* (*H. Angel*, BF) Horseshoe crab *Carcinoscorpius* sp (NHM) Fiddler crab *Uca* sp (*I. Polunin*, NHPA)

Chapter 6

58–59	Garden spider *Araneus diademata* (*J. Cooke*, OSF) Weaver ant *Oecophylla* sp (*A. Warren*, Ardea) Leaf-cutter bee *Megachile* sp (*G. du Feu*, PE) Land crab *Gecarcinus* sp (*P. Parks*, OSF) Leaf-rolling cricket *Camptonotus carolinensis* (*R. Hutchins*, MEM) Froghopper *Philaenus spumarius* (*P. Sauer*, BC) Mud-dauber wasp *Sceliphron* sp (*J. Cooke*, OSF) Leaf-rolling beetle *Apoderus coryli* (*A. Romage*, OSF) Sand wasp *Ammophila* sp (*K. Taylor*, BC)
60–61	Caddis fly (NHMPU)
62–63	Honeycomb of *Apis* sp (NHMPU)
64–65	Spiders' webs (*P. Hillyard & J. Martin*, NHM)
66	Nest of tropical termites *Cubitermes* sp (*P. Ward*, BC) Nest of compass termites *Amitermes meridionalis* (NHMPU)

Chapter 7

99
- Bluebottle *Calliphora* sp (NHMPU)
- Dust mite *Dermatophagoides farinae* (NHMPU)
- Silverfish *Lepisma saccharina* (M. Tweedie)
- Bed bug *Cimex lectularius* (S. Dalton, NHPA)
- Cockroach *Periplaneta americana* (G.I. Bernard, OSF)

101
- Devil's coach horse beetle *Staphylinus olens* (M. Tweedie)
- Common wasp *Vespula vulgaris* (NHMPU)
- Bluebottle *Calliphora* sp (NHMPU)
- Clothes moth *Tineola bisselliella* (NHMPU)
- Garden ants *Lasius niger* (NHM)
- Cat flea *Ctenocephalides felis* (S. Dalton, NHPA)

103
- Carpet beetle larva *Anthrenus verbasci* (L. Crowhurst, OSF)
- Larder or bacon beetle *Dermestes lardarius* (G.I. Bernard, OSF)
- Cheese with cheese mite *Tyrophagus casei* (NHMPU)
- Death-watch beetle *Xestobium rufovillosum* (H. Angel, BF)
- Spider beetle *Ptinus tectus* (S. Dalton, NHPA)
- Biscuit beetle *Stegobium paniceum* (NHM)

105
- Grain weevil *Sitophilus granarius* (NHMPU)
- Clover mite *Bryobia praetiosa* (NHMPU)
- Woodlice *Oniscus asellus* and *Porcellio scaber* (M. Tweedie)
- Silverfish *Lepisma saccharina* (NHMPU)
- House spider *Tegenaria gigantia* (NHMPU)

Chapter 8

107
- Anopheline mosquito *Anopheles* sp (NHMPU)

109
- Malaria parasite *Plasmodium falciparum* (Prof. W. Peters, LSHTM)
- Black fly *Simulium damnosum* (S. van Eysinga-Meredith)
- Tsetse fly *Glossina palpalis* (A. Bannister, NHPA)

111
- Desert locust *Schistocerca gregaria* (NHMPU)
- Brown planthopper *Nilaparvata lugens* (M.R. Wilson, NHM)
- Nettle caterpillar *Setora nitens* (J.D. Holloway, CABI)
- Desert locust (hoppers) *Schistocerca gregaria* (G. Popov, NHM)
- Aphids *Sipha glyceriae* (NHMPU)
- Colorado beetle *Leptinotarsa decemlineata* (R. Knightsbridge, NHPA)
- Migratory locusts (mating) *Locusta migratoria* (G. Popov, NHM)
- Armyworm *Spodoptera exempta* (M. Simmonds, BCUL)

Chapter 9

115
- Miscellaneous fossil arthropods (NHMPU)

117
- Giant water scorpion *Pterygotus anglicus* (NHMPU)
- Model giant water scorpion *Baltoeurypterus tetragonophthalmus* (NHMPU)
- Fossil horseshoe crab *Mesolimulus walchii* (NHMPU)
- Horseshoe crabs *Limulus polyphemus* (D. Attenborough)

119
- Fossil trilobites *Calymene blumenbachii* (NHMPU)
- Pill millipedes *Glomeris marginata* (M. Tweedie)

Burgess Shale fossils (NHM)

91	Fossil prawn *Acanthochirana cordata* (NHMPU) Fossil trilobite *Leonaspis coronata* (NHMPU) Fossil crab *Archaeogeryon peruvianus* (NHMPU) Fossil lobster *Hoploparis longimana* (NHMPU) Cranefly wing *Tipula* sp (NHMPU)
92	Insects in amber: two flies, *Rhagio* sp & *Chrysopilus* sp; unknown orthoptera (grasshopper) (NHMPU) Arachnids in amber: two spiders, *Phalafium* sp (harvestman) & *Oxyopes* sp (NHMPU)
93	Burgess Shale arthropod *Waptis fieldensis* (NHMPU) Burgess Shale arthropod *Marrella splendens* (NHMPU) Burgess Shale arthropod claw *Anomalocaris canadensis* (NHMPU)
94	Fossil insect *Cyclophthalmus senior* (NHMPU) Fossil scorpion *Paraisobuthus prantli* (NHMPU) Fossil scorpion *Eoscorpius sparthensis* (NHMPU)
95	Fossil dragonfly *Cymatophlebia longialata* (NHMPU)
96–97	Trilobites *Encrinurus punctatus* (NHMPU)
98	Fossil hymenoptera: *Anthorphorites titania, Polistes kiryanus, Xylocopa senilis, Odynerus* sp (?) (NHMPU) Fossil wasp *Polistes kirbyanus* (NHMPU) Fossil flower *Porana oeningensis* (NHMPU) Fossil beetle *Ancylocheira* sp (NHMPU) Fossil bee *Anthophorites titania* (NHMPU)
99	Small tortoiseshell butterfly *Aglais urticae* (NHMPU)
100–101	Velvet worm *Peripatus* sp (A. Bannister, NHPA) Sea spider *Colossendeis* sp (H. Chaumeton, NHPA) Water bear Phylum Tardigrda (M. Walker, NHPA)
102	Entomologist with specimens (NHMPU)

Abbreviations:

BC	=	Bruce Coleman
BF	=	Biofotos
BCUL	=	Birkbeck College, University of London
CABI	=	CAB International
HS	=	Holt Studios
MEM	=	Mississippi Entomological Museum
NHM	=	Natural History Museum
NHMPU	=	Natural History Museum Photo Unit
NHPA	=	Natural History Photographic Agency
OSF	=	Oxford Scientific Films
PE	=	Planet Earth Pictures
PP	=	Premaphotos
RUV	=	Radford University, Virginia

All rights reserved. No part of this publication may be reproduced, stored in a retrieval system, or transmitted, in any form or by any means, electronic, mechanical, photocopying, recording, or otherwise, without the prior permission of the publisher.

© British Museum (Natural History) 1991
Cromwell Road, London SW7 5BD.

Printed in Hong Kong

British Library Cataloguing in Publication Data

Ladybirds & Lobsters, scorpions & centipedes.

1. Arthropoda
1. British Museum (Natural History),
Department of public services
595'.2

ISBN 0-565-01084-0